REPORT

OF

T. S. HARDEE,

CITY SURVEYOR OF NEW ORLEANS,

ON A SPECIAL SURVEY

OF

LAKE PONTCHARTRAIN,

IN CONNECTION WITH THE EFFECTS OF THE

BONNET CARRE CREVASSE,

AND THE DANGERS FROM

OVERFLOW,

WITH ACCOMPANYING CHART.

PUBLISHED BY AUTHORITY OF THE CITY COUNCIL.

NEW ORLEANS:
A. W. HYATT, STATIONER AND PRINTER, 38 CAMP STREET,
1876.

LAKE PONTCHARTRAIN.

OFFICIAL REPORT OF THE CITY SURVEYOR.

The effects of Bonnet Carrè Crevasse on the physical condition of the Lake—Impending dangers of overflow from a concurrence of river floods with easterly gales—An appropriation from Congress suggested to close the Crevasse.

SURVEYOR'S OFFICE,
New Orleans, September 25, 1876.

To the Honorable Mayor and Administrators of the City of New Orleans :

GENTLEMEN--In response to the resolution adopted by your honorable body on the 2d of May, 1876, the United States Government generously placed the revenue cutter Dix at the service of the city authorities, for the purpose of assisting in a special survey of Lake Pontchartrain, and I now have the honor to report, as your engineer officer in charge, that the work has been conducted to a satisfactory termination.

The object of the survey was to determine the effect of the Bonnet Carrè crevasse on the physical condition of the lake, and to deduce from the facts thereby elicited the prospective dangers which might threaten the city of New Orleans from that source. The results are not as full and complete in some details as might have been desired, owing to the fact that there was no previous survey made by the Government, and therefore no records in existence by which to compare present soundings with those of the past. In consequence of this deficiency, and the want of means on the part of the city to undertake the amount of work involved in an original survey, attention was mainly directed to discover the existence of sedimentary deposits in the lake, and then by suitable examinations to determine the location and area of the same, believing that the information thus obtained, in connection with the knowledge of other well established physical peculiarities, would satisfy all requirements in the premises.

With the view of accomplishing the work projected, a new contrivance of a combination rod was ordered to be constructed, which answered successfully the double purpose of taking soundings and borings at the same time.

PRELIMINARY EXAMINATIONS,

made in a general way over the lake, at various points, from Pass Manchac to the Rigolets, developed the fact that the deposits were being made principally in a compact mass at the western end of the lake, where the rapid discharge of river overflow is finally checked by contact with the tidal expanse, and to this locality the labors of the survey were chiefly directed.

The present crevasse, which was caused in the spring of 1874 by a breach in the levee at Bonnet Carrè bend, about thirty-five miles above the city, is now 1370 feet in width, in a direct line across the gap, and as the discharge of water courses towards Lake Pontchartrain, five miles distant, it widens in a fan-like shape, so that by the time it reaches the shore of the lake the flow of water has attained a breadth of more than twenty-two miles, as will be seen by the section colored blue on the accompanying chart. The difference in elevation between the highest water in the river at the head of the crevasse, and the ordinary tidal level of Lake Pontchartrain is about eighteen feet, and the discharge of water through the crevasse when the river is in the neighborhood of flood height has been estimated to be about one-tenth of the volume of the entire river which passes at that point. With these facts kept clearly in mind, and viewed also in connection with the immense quantity of sediment known to be carried in the stream, it can be readily conceived what must be the results necessarily involved in the very nature of such an outpouring. A very large proportion of sediment is of course precipitated before reaching the lake shore, but enough is still held in suspension to produce marked effects when the flow of water is at last checked by debouching into the lake.

The accompanying map shows the location and

AREA OF THE DEPOSITS,

as determined by the examinations. The figures shown in the section, colored brown upon the chart, represent the thickness of the deposits at the various points where the rod was used. In recording these different borings care was taken to show only what was deemed to bear indisputable evidences of river mud or alluvion, embracing principally the results of the three successive crevasses since that of 1874, and possibly embodying those of other overflows anterior to that date, and it may be of some even before the levee system was adopted. It will be perceived that the greatest width and depth of deposit has been made opposite Frenier Station, on the Jackson Railroad, or where the main body of the crevasse water pours into the lake. At this point the cross section from the shore outward shows the deposit to have been made over a distance exceeding five miles, and the greatest thickness in that section to be about six and a half feet. The whole area of deposit embraces a compact mass, some twenty-two miles in length by an average of four miles in width, and contains, by calculation, about two hundred million cubic yards of sedimentary matter. To furnish some practical idea of the immensity of this accumulation, it may be stated, that if the material were distributed over the inhabited portion of New Orleans, it would be sufficient to raise the entire ground of the city more than four feet above its present low level.

The capacity of Lake Pontchartrain as a reservoir, its entire water contents being displaced, will allow the substitution there-for of about ten thousand million cubic yards of earth or sedimentary matter. With the knowledge, then, that in a given number of years the capacity of the lake has been reduced to the extent of what two hundred million cubic yards of sediment would displace, it is within reach of arithmetical computation to calculate the length of time it would require to fill up the body of the lake, provided the crevasse should remain permanently open, with the same average discharge of water annually, and with the same general laws governing its action in the future as in the past. But as the number of years in which this estimated deposit has been accumulating cannot be ascertained with any degree of accuracy, it follows that no deductions can be made which would fix, even approximately, the happening of such an event at any specified date or period in the future. It can safely be predicted, however, that many generations will come and go before the citizens of New Orleans would be deprived of the benefits of Lake Pontchartrain as a navigable water. Agencies more active than the slow one of shoaling must, therefore, be taken into consideration as liable to produce anything like the immediate dangers which have been apprehended from the altered physical condition of the lake.

It has been clearly established that the

EFFECT OF THE CREVASSE,

up to the present time, has been to raise considerably the ordinary level of Lake Pontchartrain, which abnormal condition lasts generally about four months in the year, or during the flood period in the Mississippi River. This elevation of the surface of the lake is almost exclusively due to the immense volume of foreign water constantly precipitated into its basin from the crevasse, which must necessarily create for itself a new plane of discharge to the Gulf by the consequent forced raising of the water.

With this great cause of danger from the large amount of fresh water, are combined the injuries which hreaten from the periodical conditions of Lake Pontchartrain by overflow caused by accumulations of sea water.

It has been contended that the obstructions caused by the bridges and earthworks of the Mobile Railroad, between the Gentilly ridge and the highlands east of Pearl River, have contributed largely to the elevation of the water line in the lake, by damming and backing up the crevasse water, thus impeding its free discharge into the Gulf. A careful investigation of the line of road, when the water was at its highest mark, did not develop any facts which warrant such an assertion. On the contrary, the construction of the railroad embankments across the lowlands or marshes between the points mentioned is deemed a positive benefit to the city of New Orleans as a prevention of overflow. It unquestionably acts as a barrier to shut out the Gulf water, which in severe easterly gales rolls towards Lake Pontchartrain in a solid mass of thirty miles in width, and of a depth sufficient to inundate simultaneously the whole intervening marsh or lowland, upon which it is precipitated in a few hours' time. This massive wave, when it comes, is now arrested by the railroad embankment, acting as a protection levee, and the supply to the lake basin can only be made through the limited openings where the bridges are constructed. Thus the influx of this immense quantity of water is rendered more gradual, and would occupy days before it would pass through these narrow inlets, during which the subsidence of the gale or a fortunate change of wind would arrest its further accumulation and increase, and tend to discharge back into the Gulf by the outflow what had already been received. It will thus be seen that, in the confinement of the

RECEPTION OF SEA WATER

to these circumscribed entrances, these inlets serve to a certain degree the purposes of locks in lessening and delaying the admission of water into the lake, when without the embankments the whole expanse of thirty miles would afford the opportunity for the rapid introduction of the whole bulk and force of the mammoth wave in a few hours. While it is plain, therefore, that the embankments act as barriers, and the openings as embouchures regulate the admission of water, it is equally evident, that if by a series of easterly gales, continuing through a considerable length of time, the lake should receive the maximum of water, and thereby overflow the city from the rear; the same reasoning shows that these narrow outlets would retard the discharge back into the Gulf, delay relief from the extraneous surplus of foreign water, and necessarily prolong the inundation. The effect is best illustrated by a comparison of the lake to the body, while the openings would correspond to the narrow neck of a bottle, the smaller the latter is, the slower the process is in filling the vessel, and when once filled, the emptying of its contents, for the same reason, must be correspondingly gradual. The latter contingency could arise, however, only by the occurrence of a continued series of violent easterly gales to an extent hitherto unknown, and happening in rapid succession, before the discharge could be effected. Such a state of things could scarcely be anticipated, and at the worst the damage would have already been inflicted, and would only be slightly aggravated by a somewhat longer continuance of the overflow.

In view of this theory, therefore, which is believed to be correct, the apparent obstruction of the railroad is converted into an absolute advantage, for the great damages by back overflow have always been produced by the quick and sudden precipitation of this Gulf water into the lake basin, and its consequent encroachments upon the contiguous lowlands in the rear.

THE RECORDS

in this office show a slight annual increase in the height of the water in the lake for the past three years during the flood period of the crevasse, and before any examinations were made, it was presumed that the shoaling, which was believed to be progressing more extensively than has been found to be the case, might produce such effects. But the variation being only a few inches between the seasons of 1874 and 1876, it is concluded that other causes, such as the differences in the actual volumes of water passing through, together with the action of the winds, etc., might have induced these very changes and irregularities.

It is considered, therefore, that the question of increased danger to New Orleans depends solely, upon the presence of so large a quantity of crevasse water in the lake, and the consequent diminution of its normal capacity as a reservoir during high water, thus limiting its ability to receive and hold the immense volume of additional water which might be blown into it from the Gulf during the prevalence of easterly gales. But when the two conditions concur it is evident that the danger becomes intensified, because the incoming storm wave from the Gulf would completely hold in check the outgoing flood of crevasse water, and with the happening of such a coincidence in an aggravated degree, it can be easily conceived how New Orleans might be subjected to the severest and most damaging overflow she has ever experienced. Such a calamity was barely escaped last spring, when the water in the lake and in the several canal basins was higher than ever before known. It only required at that time a continuance of the wind storm for a few hours longer to have so raised the water that it would have passed bodily over the tops of the low protection levees in the rear, and even been forced over the higher natural barrier of Metairie Ridge, thus inundating the entire city from the river to the lake.

Such being the

CONDITION OF AFFAIRS,

it was deemed pertinent to the purposes of the examination to consider and digest the best plan to be adopted which might mitigate or prevent some of the evil consequences of the situation, even if it did not entirely establish perfect security from the dangers threatened. The best means practicable, which suggested themselves, was the closing of the crevasse, in order to shut out the river water, as one of the greatest disturbing causes of the normal condition of the lake. Indeed, the immense expense which would accompany any attempt to preclude the sea water from encroachment, debarred any effort in that direction, and necessitated the turning of our attention to the control of the river water, in order to establish comparative safety, relying on the completion of the back protection levee in the future as the only measure of complete and permanent security to the city. In view, therefore, of what has been adduced in this discussion, it is clear that the interests of New Orleans are deeply involved in the closing of the Bonnet Carre crevasse, for with this accomplished a great factor in the problem of danger is at once eliminated. It is all important also, that the plans for the accomplishment of the work should be matured without delay, in order that it may be undertaken in time to control the Mississippi before the recurrence of another flood.

It is estimated that $200,000 would be required to re-establish the security of the levee front in the vicinity of the crevasse, and incidental to it, an additional $50,000 would be needed for the purpose of repairing the levees all the way below, which have been neglected for several years past, on account of the flood-line in the river being decreased, as an effect of the crevasse above, and the fears of those interested in planting proportionately removed. A total expenditure, therefore, of $250,000 is deemed to be sufficient to meet both demands of the situation, and to satisfy all interests concerned.

In the present

EMBARRASSED FINANCIAL CONDITION

of the city and State, there is scarcely a reasonable hope that the required amount of money can be obtained from either of those sources, and reliance, therefore, must be placed upon the strong arm of the General Government to afford the relief so much needed. It is suggested, also, that a community of interest exists on this subject between the seacoast of Mississippi and the city of New Orleans, and that concert of action between the two, in preparing a memorial to Congress for aid in this particular work, is the plan which should be adopted, and which promises a measure of success. The whole coast of Mississippi, from Pearl River to Pascagoula, embracing millions of dollars of improved property, is offended with the presence of fresh water from the crevasse at a season of the year when it is a resort for health and pleasure, and thus the material interests of that section of the State are very seriously damaged. Mississippi certainly has some claims upon the Government for the improvement of her seacoast, as she is, perhaps, the only State bordering on the ocean waters which does not annually receive appropriations for that object, and although the expenditure in closing the Bonnet Carre crevasse would not be applied immediately to the coast front, yet the advantages to flow from it for her benefit would be no less direct and positive. Her representatives in Congress would no doubt join in exertions with those from Louisiana, and use every means in their power to obtain the necessary appropriation at the earliest possible moment. If a bill could be matured and presented among the first days of the ensuing session in December, it is possible to have it acted upon in time to enjoy the benefits of an appropriation for the closing of the crevasse before the coming of the next flood, otherwise the dangers and damages from the situation must be endured for another year.

It affords me pleasure, on behalf of the city, to acknowledge the favors extended by the officers of the United States Customs Department here, in their uniform courtesy and for the facilities afforded me in the prosecution of the work; also, for the hospitalities extended to the surveying party by the officers of the revenue cutter Dix, as well as for the valuable assistance rendered so willingly by them and their crew.

My thanks are also eminently due to Messrs. Harrod and Reynolds, my assistants, and the detailed party of employees, who so efficiently labored and co-operated to bring the work to a satisfactory conclusion.

Very respectfully,

THOMAS S. HARDEE,

City Surveyor.

www.ingramcontent.com/pod-product-compliance
Lightning Source LLC
LaVergne TN
LVHW020642110826
845149LV00004B/1320

* 9 7 8 1 4 1 8 1 8 9 6 6 2 *